MW01286130

Forever Grateful To You: